Rescuing Baby Rabbits & Hares

Guide For Wildlife Rehabilitators

Cottontails, Hares, & Jackrabbits

Paperback ISBN 979-8-9991174-6-5
eBook ISBN 979-8-9991174-7-2

Thank you
For Your Support

Fox Run Environmental Education Center
is a 501c3 Non-Profit.

Environmental Education
Organic Gardening
Wildlife Conservation and Rehabilitation

Ame Vanorio, Founder

Table of Contents

Marsh Rabbit
Photo credit: Richard Stovall

Table of Contents

Introduction

Welcome to the wonderful world of wildlife rescue! If you love wildlife and want to help them this book is for you!

This book is part of a wildlife rehabilitation series that consists of individual books that cover specific species. The first book is Squirrels. Each book has step-by-step instructions for that species. If you want to know how to become a wildlife rehabilitator then check out my book **"Getting Started With Wildlife Rehabilitation"** also available on Amazon in paperback and Kindle.

Please note that some topics such as pests and diseases affect many species and will occur in several books, however, I focus on symptoms and treatments that are species-specific.

Rabbits live all over the world. This guide focuses on cottontails and hares that live in North American. They are similar in terms of care. I will use the term 'rabbit' as a collective term except when I want to point out differences with hares.

This book takes a **deep dive specifically into cottontails with some additional information on jackrabbits and snowshoe hares**. What are the first steps we need to take when a baby rabbit has come into our care? What pests and diseases are they prone to and how to treat and prevent them? How should you feed an infant rabbit? How can I raise a rabbit so it is prepared to be released back into the wild? Why knowing their life cycle and natural history is so important.

Legal Statement

Your state (or country) may have laws pertaining to rehabilitating rabbits and you should comply with those laws for the best care of the animals.

Each state has its own guidelines for becoming a wildlife rehabilitator, what species it covers, and providing care for infant wildlife. These are typically listed under the state's Department of Fish and Wildlife or DNR.

My wildlife rehabilitation and conservation career and experience have taken place in the USA. If you are in another country what's a common disease or animal for me may not be as common for you and vice versa.

My books are meant to help you recognize and treat common problems that present themselves with animals in your care. It should in **no way replace advice given by your veterinarian**. I assume that you are working with a qualified veterinarian to provide the best possible outcomes. I also recommend that new wildlife rehabilitators spend some time working with an experienced rehabber to "learn the ropes".

One of my rescues
Eastern Cottontail
Photo by Ame Vanorio

Why?

I often get asked why I rehabilitate rabbits (besides the fact that they are cute). After all, rabbits are seemingly everywhere, even in city neighborhoods. Urban biodiversity matters. Rabbits help disperse seeds, graze vegetation, and support predators that control rodent populations.

Rehabilitation not only helps individual animals but also supports the broader ecosystem. By rehabilitating rabbits, we contribute to the well-being of habitats and enhance our local biodiversity.

Because rabbits depend on dense cover and native vegetation, their presence can signal habitat quality. When local rabbit populations decline, it often reflects changes in land use, mowing frequency, predator imbalance, or pesticide use.

Yes, rabbits are important prey animals. They are a natural part of the food web and serve as a primary food source for coyotes, foxes, owls, and bobcats to name a few. In some ecosystems, especially northern forests and tundra, hare population cycles directly influence predator survival.

That can be something that rehabilitators struggle with. I know I do! Any animal that we have cared for we want to have a long happy life. Knowing "your baby" may be someone else's dinner can be hard.

Many times our rehab animals come to us through human error. Mowing grass in backyards or farm fields may destroy nests or injure babies that are not old enough to run away. Domestic cats and dogs hunt rabbits. Habitat fragmentation pushes rabbits into roads and suburban spaces. Development removes the dense cover they depend on.

Wildlife rehabilitation centers often see trends before the public does. A spike in lawnmower injuries in spring tells you something about nesting timing and mowing practices. Increased cases of emaciation may signal drought or habitat stress. Rescue work becomes a data point. It informs land management, public education, and conservation planning.

Every rabbit brought to a wildlife center is an opportunity to teach. People learn that mother rabbits do not stay at the nest all day. They learn why outdoor cats are a major predator. They learn how simple changes such as waiting to mow, checking brush piles, and planting native shrubs, can protect wildlife.

Prevention is always better than rescue. Rehabilitation programs often reduce future intakes by teaching coexistence. That education has long-term impact beyond one animal.

For organizations focused on environmental education, this is powerful. Rabbits are familiar, visible wildlife. They make strong ambassadors for habitat awareness. Educating the public about rabbit behavior and needs can reduce human-wildlife conflicts. Understanding them can lead to more harmonious coexistence in urban and suburban areas.

As wildlife rehabilitators we promote a compassionate approach to interacting with wildlife. We teach our fellow humans the importance of wildlife in our backyards and beyond and humane ways to deal with conflicts.

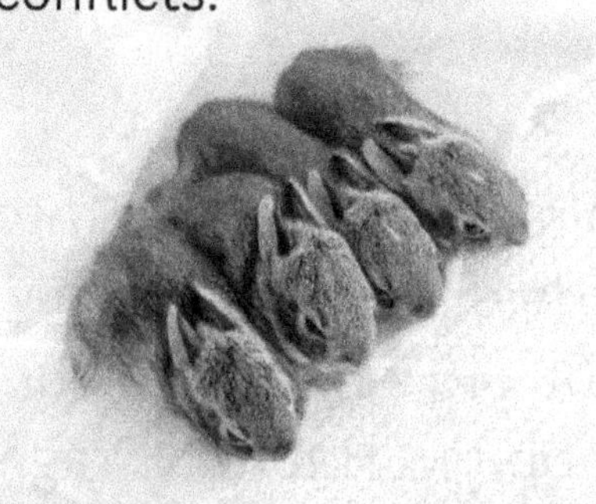

Eastern
Cottontail
kits

Common Reasons for Rabbit Intervention

Vehicle Collisions:
Rabbits get hit by cars while crossing roads. Depending on food sources the Eastern cottontail ranges 1 - 15 acres or up to a mile from their birth place. The black-tailed jackrabbit feeds within a home range up to 475 acres (192 hectares).

Predator Attacks:
Injuries from attacks by domestic pets (cats and dogs), birds of prey, or other wild predators. Cat bites especially need medical intervention.

Orphaned or Abandoned
The mother may have been killed by a predator, vehicle, or other incidents, leaving babies orphaned.

Nest Disturbance:
Human activities such as mowing and other yard work can destroy nests, causing babies to be separated from their mothers.

Disease
Mange: Caused by mites, leading to hair loss and skin infections.
Viral or Bacterial Infections:

Poisoning
Pesticides and Rodenticides: Accidental ingestion of chemicals used in gardens or around homes. Including pesticides and traps used for mice.

Why Are Rabbits So Difficult?

Rabbits and hares have a reputation for being difficult. There are many reasons for this. Stress and a unique digestive system are issues you need to be aware of. Here is a quick overview and we will talk about both of these in more detail in the sections on Anxiety and Feeding.

Stress

Wildlife rehabilitation is inherently stressful for all wild animals. That matters a lot for baby rabbits because they are prey animals, don't adapt well to captivity, and can decline quickly. One morning they seem to be fine and adjusting well and that evening they have diarrhea or bloat.

Human contact and handling are major stressors for prey species in captivity, which is why rehab standards push "quiet, dark, low-traffic, minimize handling."

Digestion

Rabbits have an interesting and complex digestive system. Infants are born with little microfauna in their intestines. That's the good stuff that helps us break down foods. Plant fiber is difficult for the system to break down. They start receiving microbes from mom and if mom is out of the picture this can cause an imbalance in the gut. In addition, it causes them to be more vulnerable to pathogens such as bad bacteria or viruses.

How To Tell If A Baby Rabbit Needs Intervention?

Baby cottontails are often misidentified as orphans! They mature very quickly. Make sure to thoroughly study their life cycle before accepting one into rehabilitation.

Nests are often disturbed by yard work. Or pets and children may discover a nest. Have them move away from the nest. If babies are uninjured gently repair the nest if necessary. Teach your audience that rabbit and hare moms do not stay in the nest. They visit in order to nurse.

Teach finders the string method. After covering the nest with the nesting material lay two or three strings in an X. If the mother comes back to nurse she will inadvertently disturb the string. If the strings are not disturbed in 24 hours then yes, the rabbits need intervention.

Older babies naturally go through a stage where they freeze and don't move. Teach finders this is normal and to not disturb them.

Questions to ask of the finder.
- If you find a wild baby, look around to see if you see mom. We never want to create an orphan. A wild mom is the best possible mother for the baby.
- Does the nest have blood or injured babies in it?
- Does the baby have fur?
- Is the baby eating grass?

Weaned cottontail

- Dehydration over 5% or emaciation
- Covered in parasites such as fleas, ticks, and/or lice
- Matted or dirty fur
- Wounds with bleeding, exposed organs, or maggots.
- Possible concussion.
- Breathing problems
- In shock or unconscious
- Cold, low body temperature

Aging a Baby Rabbit

Cottontails (Sylvilagus species) are altricial, which means they are born blind, hairless, and helpless. They depend completely on their mother and stay hidden in a fur-lined nest for the first weeks of life. This nesting strategy helps protect them in dense vegetation.

They are born with a white stripe on their forehead. **This stripe is not a sign of age!** The stripe may not disappear as the rabbit ages or it may fade in a couple of weeks. Each animal is different.

Operation Wildlife has a good rabbit chart and Moon Gazers has a comparison of rabbit and hare babies.

One to Two Days
Born naked, with greyish skin and pink nose and feet. Eyes are closed. The ears look like they are pinned back. Can produce an ear-piercing scream with mouths wide open when scared.

A velvety black fur starts to grow. Fur begins to come in around 3-4 days and continues until the baby is fully furred at about 3 weeks.They typically have a tiny white spot (blaze) on their foreheads however its not a reliable age marker. Eyes are closed and ears are still flat to the head.

11 to 22 Days
Babies are starting to explore on their own at 11 - 22 days. Typically by two weeks, they start leaving the nest to nibble on plants, but will continue to nurse.

21 - 28 days old
They will fit in the palm of your hand. Their fur becomes more fluffy, ears are upright and open - they startle easily. .

Aging a Baby Rabbit cont

At 4 weeks they are ready to be independent. For release they should be four to five inches long and about the size of a tennis ball.

Male desert cottontail at 8 weeks, and the same cottontail at 16 months of age
Photo credit: Jessie Eastland

Juvenile swamp rabbit
Photo credit:
Brandon Johnson

Cottontail Developmental Chart

Day 1 - 3	They have no fur, eyes and ears are closed. Need urine stimulation.	Formula feed 3 times per day. Every 4 hours during the day. Average weight is 15 - 25 grams
Day 4-7	Fur begins to grow	Formula feed 2 - 3 times per day depending on how readily they eat. Weight @ 24 - 35 grams
Day 7 - 10	Eyes and ears open.	Formula feed 2X a day. start to offer small amounts of greens & hay for them to try. Give probiotics. Weight averages 35 - 50 grams.
Day 11 - 14	Starting to hop around enclosure. Starting to urinate on own. Offer a variety of green foods. They should be eager to sample them. Monitor weight daily as they transition to solids.	Feed formula once a day if they are eating enough greens. Otherwise stay with 2x. They may reject being fed. Offer formula in a shallow dish and remove agter 30 minutes. Weight 45 - 60 grams
Day 14 - 21	Fully furred and hopping around. Stimulate till they poo & pee on own which is around this time.	Eating greens and hay - available at all times. Continue probiotics. Weight @70 - 95 grams.
Day 22 - 28	Start to show fear and hide from you . Move ears to listen and are alert.	Give more space to prepare for release. Always have timothy hay available. Give greens. Weigh 100 - 200 grams

Aging A Baby Hare

Hare young (Lepus species) are precocial, which means they are born fully furred, with open eyes, and able to move within hours. Instead of being raised in a covered nest, they are born in shallow depressions called forms. Hare newborns will start to nibble on solids and grasses but still need formula.

Key Age Indicators for Leverets

0–3 Days: Born with fur and eyes open, often found in shallow, open, above-ground nests (forms).
1–2 Weeks: They are fully mobile, can hop, and start nibbling on solid food. Should have hay available at all times.
3–4 Weeks: They begin leaving the nest, with weaning starting around the 4th week.
5+ Weeks: Needs outdoor enclosure. They become more independent, and in rehab, they show signs of "wildness" (crawling up walls, resisting handling).

Snowshoe hare kit - @4 days. Born they are fully furred, their eyes are open and they are capable of hopping around almost immediately. Leverets nurse only once a day, usually in the evening however in rehab they may need to be fed more often to make sure the get the quanity needed.

A new born hare is about the same size as a 3-4 week old cottontail.

Reuniting

Sometimes nests are disturbed when people are mowing the yard or kids and pets are playing. There are times when baby rabbits can be returned to the nest. You can gently rebuild it. Wear gloves. Place the babies back together in the original nest spot. Cover them lightly with the original nesting material if available. If not, use dry grass. Do not add excessive material. The nest is usually shallow and simple.

Mother rabbits will not reject babies because of human scent. That is a myth. Place two pieces of yarn or string in a loose X or star pattern over the nest. Leave the area completely undisturbed overnight.

Check early the next morning. If the string has been moved and the babies look fed (round bellies, warm, quiet), the mother returned.
If the string is untouched and the babies look weak or thin they need intervention.

Keep children, dogs and outdoor cats away from the area. If possible, create a simple barrier such as an upside-down laundry basket weighted at the corners with a small gap for the mother to enter and exit.

Older babies will often freeze when they are scared and finders will walk up to them assuming their is a problem. Teach people this is a normal reaction and to give the rabbits space.

Kits
in a
nest

Make A Wildlife Rescue Car Kit

- Safety vest
- Flashlight
- Leather gloves
- Goggles
- Cat carrier
- Towels
- Piece of cardboard
- Hand warmers
- Wire cutters

Place everything in the cat carrier or a sturdy box and tuck into the back of your car

Be Safe! Always make sure that you pull over in a safe spot and consider traffic when checking on the animal. Put on your emergency blinkers.

Important Numbers

Print out a list of your states licensed wildlife rehabilitators

You can watch my video on making a wildlife rescue car kit. @foxruneec

First Steps:

- Heat
- Hydration
- Formula Feeding

I will get into more species specific information below. But this advice applies to any baby rabbit or hare that comes into your care.
ALL incoming babies need heat, hydration, and then after at least 12 hours, food.

Heat

The first step is to gradually warm the baby rabbit. Infant rabbits have some ability to thermoregulate at birth, relying heavily on huddling together and residing in a warm, fur-lined nest to maintain body temperature during their first 10–14 days. While they show some behavioral and physiological responses within hours of birth, they are highly susceptible to exposure. Newborns are hairless and more vulnerable.

An older baby with open eyes is capable of thermoregulating however due to shock and/or trauma their body may not be providing necessary heat.

Rabbits cannot sweat and are prone to heat stress so they should be closely monitored during the warming process.

Important Tips:
- Heat only under half the enclosure
- Always include a towel barrier
- Never place the animal directly on a heating pad

Heat

Caution is needed when giving supplemental heat. A neonate or other animal that can't move should never be placed directly on the heat source. They may become overheated and unable to escape.

In addition, infant rabbits and hares are sensitive to too much heat. Keep the heating pad on low and never place them directly on the pad. A hot water heating bottle in the cage/tank may also work well.

Siblings should be kept together and they will snuggle and help keep each other warm.

At core body temperatures below 94°F (34.4°C), gastrointestinalileus and bradycardia occur, and oral fluids should not be given until the patient is warmed. Rapid warming is discouraged in compromised neonates.
Hypothermic neonates should be rewarmed slowly and carefully to avoid complications.

North American hares (genus Lepus), such as the snowshoe hares and jackrabbits, are born precocial. They are fully furred, with open eyes, and capable of movement shortly after birth—enabling them to begin maintaining their body temperature immediately.

Hares are capable of thermoregulation from day one, even in cold conditions, though they rely on huddling for better survival.

Average Temperatures of Rabbits and Hares

Across lagomorphs (rabbits and hares), normal core temperature generally falls between 101.3–104 °F and 38.5–40.0 °C under resting conditions.

Most veterinary references list a normal rectal temperature of domestic rabbits at **38.5–40.0 °C (101.3–104.0 °F)**. Studies of wild rabbits and hares show the same range.
Source Merck and Carpenter's Exotic Animal Formulary (see resources)

In practice, wildlife rehabilitators consider a rabbit hypothermic below ~38 °C (100.4 °F) and hyperthermic above ~40.5 °C (104.9 °F), though thresholds vary slightly by situation.

Hares (genus Lepus)
Studies on hares show similar but sometimes slightly higher resting body temperatures compared to domestic rabbits.

Resting body temperature of **snowshoe hares** at approximately 39.5–40.1 °C, depending on season. (Sheriff et al., 2009)

Black-tailed jackrabbit (Lepus californicus)
Field physiology research reports resting core temperatures around 39–40 °C, with the ability to tolerate higher peripheral temperatures during desert heat exposure

Black-tailed
Jackrabbit

Keep a thermometer in the carrier or cage to monitor the air temperature. For a baby who is mobile, place a heating pad on low under one-half of the carrier. Then the baby has the option of crawling on or off the heated area.

The heating pad is the most commonly used way to offer heat. They are relatively inexpensive and easy to find. When you purchase a heating pad make sure you get on without an automatic shut-off. More and more have these as a safety method but it makes it very inconvenient for those of us who need to warm wildlife 24/7.

I like the Marunda brand just because it's made for pets and has a couple of nice features such as chew-resistant cords and a wipe-off surface. Rabbits, even small ones, like to chew so make sure you are checking on any heating pad that is inside the enclosure.

As your babies grow you can place the heating pad under the enclosure so they can not chew on it.

Heating pads also need to be monitored to make sure they are putting out consistent heat. You will also need to make sure the baby is not too hot or too cold.

Stress alone can elevate body temperature quickly. Minimize handling time and stress to reduce risk of capture-related complications.

Always use common sense. Place your own hand in the enclosure and make sure the baby is warm but not too hot!

Eastern
Cottontail
in
incubator

Author, Ame Vanorio, with a new incubator. I raised the funds through Baby Warm an online fundraising platform for wildlife rehabilitators.

Infant opossums
in my incubator

Rabbit and Hare Fun Facts

They do not dig deep burrows

Unlike European rabbits, most North American cottontails do not dig complex burrow systems. They rely on shallow ground nests and thick vegetation for cover. In winter, they may use brush piles or abandoned burrows made by other animals.

They can see nearly all the way around their body

Cottontails have eyes positioned on the sides of their head. This gives them almost 360-degree vision. The only small blind spot is directly in front of their nose.

Snowshoe hares have oversized feet

The snowshoe hare's large hind feet act like natural snowshoes. They spread the animal's weight across snow. This allows them to move efficiently in deep winter conditions.

Rehydration

Dehydration can kill a baby much faster than lack of food. Baby animals should never be offered food until after they are rehydrated. Food, and formula is food, can actually cause their system to shut down.

 I give everyone fluids to start with. Rehydrate babies for 12 hours and for at least 3 'feedings'.

The process of rehydration begins after warming the baby for thirty to sixty minutes. The most common source of dehydration that we see in babies is from lack of milk. Whether mom is dead, sick, or injured the baby is unable to nurse. However, things such as hot weather, no water source, diarrhea from illness, or blood loss can also cause dehydration.

Dehydration signs:

- poor skin elasticity
- shrunken appearance especially around eyes
- lethargy
- dry mucus membranes
- fast but weak pulse
- dark urine

We often use "tenting" to check for dehydration and skin turgor. Dehydrated skin does not bounce back into place. Its stiff and remains in a tent shape for a period of time depending on the severity of the fluid loss. This is best done along the back, at the base of the neck, and in between the shoulders. Pull the skin gently upwards and see if it bounces back into shape quickly.

Some general times with tenting:

Skin bounces back under one second is good but often we give fluids as a precaution. If it takes 2 - 3 seconds there is some dehydration, typically over 5%. Over 3 seconds and dehydration is severe, over 8%, and you likely notice shrunken skin around eyes.

Steps To Rehydration

Step One - Weight

Weight in grams. You may have done this during intake. Fluid therapy is based on weight and then calculated. (Weight is also used to measure formula and growth so make sure to have a gram scale)

Step Two - Calculate

Take their weight in grams and multiply that number by 0.1 which is 10%. That gives you their average amount of fluids needed for maintenance.

To calculate in their needs for rehydration, take their weight in grams and the percentage of dehydration you feel they have based on the information above.

You will add the two amounts together to determine how many mL of fluids the baby needs.

Example:

Weight is 40 grams. Estimated Dehydration is 4 seconds or 8%

Weight = 40 g = 0.04 kg
If estimated at 8% dehydrated:
Deficit:
 0.04 × 0.08 × 1,000 = 3.2 mL
Maintenance:
 0.04 × 100 mL/kg/day = 4 mL
Total fluids first 24 hours:
 3.2 + 4 = 7.2 mL
Divide into small frequent doses if you are administering orally.

Note: 1 cc = 1 ml. A 1 cc syringe is equivalent to a 1 mL syringe

Methods To Rehydrate

Oral

There are two main methods you can use to rehydrate. The method you use will depend on the health of the animal and the type of fluid you are using. The first is oral. This method works well if the baby is **conscious and alert**.

I typically go with an unflavored Pedialyte. This is commonly sold in drug stores and groceries. Fox Valley also carries an Electrolyte Replacer for Rehydration. You can also use Lactated Ringers orally.

Some rehabbers don't like Pedialyte because of its high salt level which they say is bad for infant rabbits.
Do not use sports drinks - they contain too much sugar and salt for wildlife.

Some rehabilitators advise not using any products with salt for cottontails. Instead use a mix of de-chlorinated water with 15% Dextrose which can be found in farm supply stores.

Using a syringe and a nipple you can give warm fluids via the mouth. For young babies, my nipples of choice are the Miracle Nipples.
Very slowly drip some fluids into the mouth. Watch for the baby to swallow. Hold the syringe upwards at a 45 degree angle. (see pictures on next page)

Don't use the oral method if the animal is unconscious, has neurological symptoms, or is having seizures.

The baby must be warm before you can provide hydration.
Do not rehydrate a cold baby.

If the baby is conscious and alert the best method is to use a syringe and gently feed the fluids. Note the syringe at a 45° angle.
Stock up on 1ml and 3ml syringes.

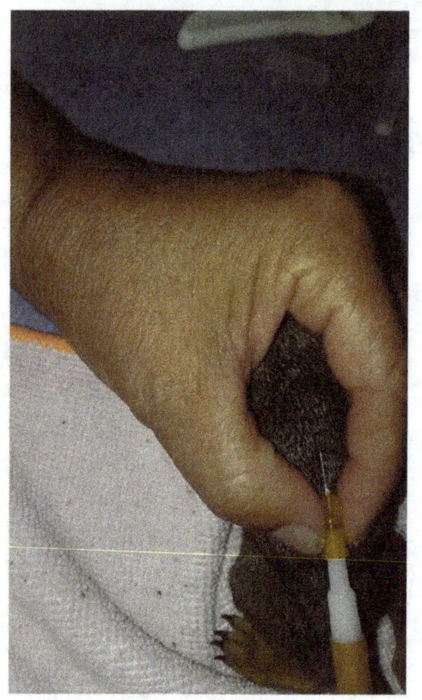

If the baby is unconscious, uncooperative, or very dehydrated the best method is sub Q - inserting a needle under the skin and allow fluids to drip in. When the fluid builds up under the skin (marble size) stop giving fluids, let the animals body absorb the fluid, and if needed administer fluids again.

Photos by Ame Vanorio
Eastren Grey Squirrel - sorry I didnt have rabbit ones. Due to their stress levels I try to work quickly when handleling.

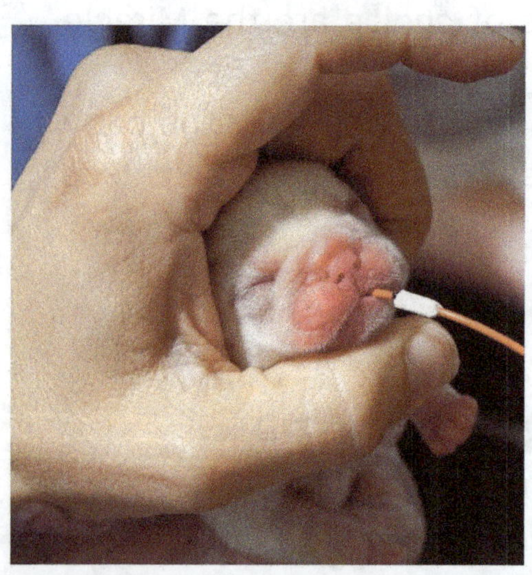

Photo by Den Guy

Tube feeding is another method where a tube is sent down the esophagus to stomach chamber. You should get trained by your veterinarian since this method can be medically tricky. I have only used it with opossums and find that rabbits typically don't need tube feeding. (Yes, that is a puppy)

Sub Q
Under the Skin

Subcutaneous means under the skin and is often referred to as SubQ. If an animal can not or will not take fluids orally then the subcutaneous method is the right option. This method is not quite as fast-acting because the fluids need to be absorbed by the body.

A needle is inserted just under the skin and the fluid is inserted via a syringe or a drip line. This is an easy technique to learn from a veterinarian or experienced wildlife rehabilitator.

For SubQ you can't use Pedialyte. You need specially formulated solutions. Lactated Ringers is the brand most often used by veterinarians. It is an isotonic solution which means it will not draw fluid from the cells via osmosis. IV solutions require a prescription. Your veterinarian should be able to sell you some and sometimes human doctors donate expired bags to wildlife rehbbers.

You can also purchase from Chewy or KVSupply with a prescription from your vet. Plan on having one to three bags on hand depending on how many animals you take in.

It's important to keep in mind that this process can cause the animal discomfort and even pain. They may cry or make sounds of distress. Their skin is dry and tight and by giving sub Q fluids we are stretching the skin. Be sensitive and know that hydration will help!

Always use warm but never hot fluids. **Never microwave fluids** as it can cause pockets of high heat within the fluids.

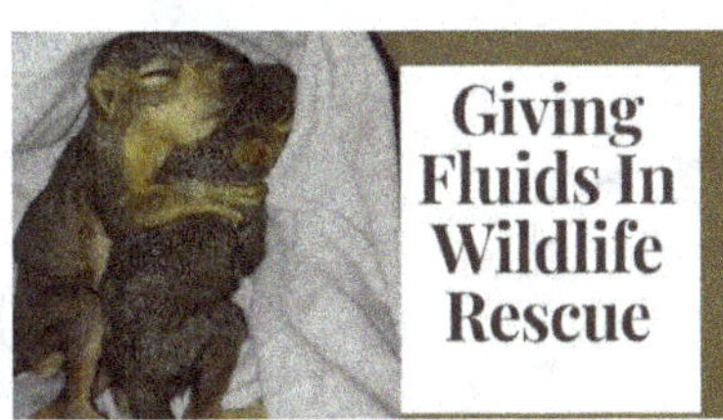

I have a video on my YouTube channel on giving fluids
https://youtu.be/LYGeH2D4L3c

29

Rabbit Digestive System

Rabbits and hares have a unique digestive system that we need to talk about first. This digestive complexity is one of the reasons that makes rabbits and hares harder to rehabilitate.

These herbivores survive on low-calorie, high-fiber plants. Their digestive system lets them extract maximum nutrition from grasses and forbs. They use hindgut fermentation. That means most of its fiber digestion happens in the large intestine, not the stomach. Sudden diet changes, low fiber intake, or stress can cause gut shutdown.

Steps of Digestion: Weaned Babies - Adults

Mouth
Like other mammals rabbit digestion begins in the mouth where food is chewed and mixed with salvia. Proper chewing reduces particle size so microbes can later ferment the material. Rabbits have a specialized teeth called Peg Teeth. These secondary incisors are located directly behind the top incisors, and are unique to lagomorphs.

Esophagus
The esophagus serves as a transport duct between the mouth and the stomach and has little effect on digestion.

Black-tailed Jackrabbit Juvenile
Photo credit: Jessie Eastland

Eastern cottontails

Stomach

Rabbits have a simple, single-chambered stomach. Acid and enzymes begin protein digestion here. Rabbits cannot vomit because of a strong cardiac sphincter, which makes them highly dependent on continuous gut movement. The rabbits movement helps to squeeze the stomach and aid in digestion. (Davies)

Small Intestine

In the small intestine, digestible nutrients such as simple sugars, amino acids, and fats are absorbed into the bloodstream. However, structural plant fiber such as cellulose cannot be broken down by normal enzymes. That fiber passes to the cecum.

The Cecum: The Fermentation Chamber
The cecum is the most important organ in rabbit digestion. It is a large blind sac that houses billions of anaerobic microbes. These bacteria ferment fiber and produce:
- Volatile fatty acids (VFAs), which provide energy
- B vitamins
- Vitamin K
- Microbial protein

Volatile fatty acids produced in the cecum are absorbed through the intestinal wall and serve as a major energy source in rabbits (Zhao et al., 2024)

Cecotrophy (Re-Ingestion of Cecotropes)
After fermentation, the cecum forms soft, nutrient-rich pellets called cecotropes. These are different from the dry fecal pellets seen in enclosures or backyards.
Rabbits consume cecotropes directly from the anus. This allows microbial protein, vitamins, and amino acids formed in the cecum to pass through the small intestine a second time for absorption.
This behavior is normal and necessary. Failure to produce or consume cecotropes is a warning sign in rehabilitation.

Sometimes wildlife rehabilitator keep a domestic rabbit or two and feed their cecotropes to their rehab patients. Babies will eat these naturally when they are ready. If availble just leave them near the nest site.

Colon and Hard Fecal Pellets
The colon plays a sorting role. Larger indigestible fiber particles move through and form dry, round fecal pellets. Smaller fermentable particles are retained or redirected toward the cecum.

Poop Guide

Normal poop is hard, dry, round pellets that crumble easily. About the size of a pea.

Cecotropes -small, soft, wet, shiny pellets, coated with mucus . They are sticky and have a pungent odor. Begin producing and consuming their own cecotropes around 3 to 4 weeks of age as they transition to solid food.

Statis poop is small/hard droppings, that occur when the rabbit is experiencing GI Stasis, a life-threatening condition where the gut stops moving. Suggests dehydration or stress and can be fatal.

Stomach Function in Baby Rabbits Before and During Weaning

So, back to our rehab babies. In the wild babies receive microorganisms via moms milk. These enzymes and fatty acids protect the infant and help with digestion

While nursing, a kit's stomach is not very acidic. The stomach pH is about 5 to 6.5, which is much less acidic than an adult rabbit. When the kit drinks milk, the milk forms a thick, soft curd in the stomach. This curd stays there for almost a full day, slowly moving into the small intestine between feedings.

The curd forms because of a special enzyme similar to rennin (enzyme). In most animals, keeping milk in the stomach that long would allow bacteria to grow. But baby rabbits have protection. When the kit digests its mother's milk, its enzymes break down certain milk fats into antimicrobial fatty acids called octanoic and decanoic acids. This substance is often called "stomach oil" or "milk oil." It helps kill harmful bacteria in the stomach.

Baby rabbits also receive antibodies from their mother. Some pass through the placenta before birth. More are received in the first milk (colostrum) after birth. These maternal antibodies help protect the kit from infection.

Kits that are fed milk replacer or milk from another species do not produce the same protective "stomach oil." As a result, hand-reared kits are much more likely to develop severe bacterial enteritis.

Infant rabbit and hare kits require a minimum of 12.5 % protein and 18% fat.

Formula Feeding

Three very important notes:

1. Never give baby formula until they have been properly rehydrated
2. Formula is food. Yes, it's a liquid but it contains nutrients, fats, and protein that must be digested.
3. The transition to formula must be slow so the body can adapt.

Transitioning From Fluids To Formula

Once the Fox Valley formula arrives, (or you have acquired <u>temporary</u> kitten formula) gently transition from the fluids to formula over six feedings. This helps the baby to slowly transition to formula without stressing its immature and stressed digestive system.

First two feedings: Combine two parts hydration fluid with one part formula (2:1)

Second two feedings: Combine one part hydration mixture to one part formula (1:1)

Last two feedings: Combine one part fluids to two parts formula (1:2)

After that, you can move on to regular strength formula feedings. If you start with KMR and need to move to Fox Valley you should also transition slowly mixing the two formulas as above.

Transition to Solid Food

Rabbits and hares grow quickly. As kits grow, they slowly shift from milk to solid food.

Around 10 days of age, in nature, they begin eating some of their mother's cecotropes. Cecotropes are coated in mucus. This coating protects the beneficial bacteria inside from being destroyed by stomach conditions. The bacteria can then pass into the intestine and begin colonizing the developing hindgut.

At 10 days start offering weaning foods. By 20 days, solid food makes up most of their diet, and normal cecotrophy has begun. By 30 days, milk intake is very low, and cecotrophy is fully established.

Changes in the Stomach During Weaning

During this same period, the production of "stomach oil" decreases. At the same time, the stomach becomes much more acidic. The pH drops to about 1 to 2, which is the adult level.

This high acidity now acts as the main defense against harmful microbes entering the stomach and small intestine.

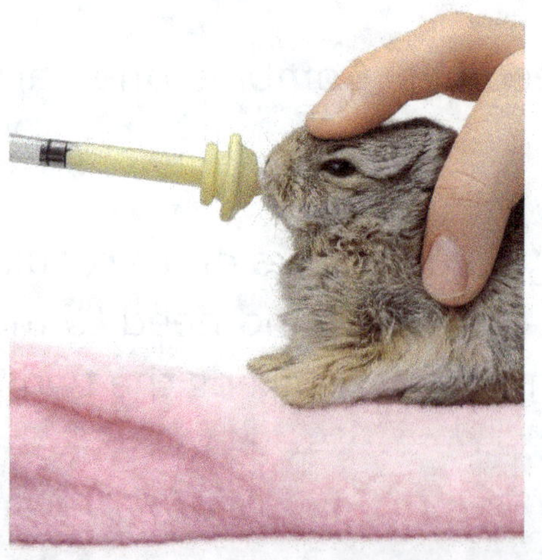

Rabbits who can sit up should lay horizontal for feeding.

What I recommend

Fox Valley Animal Nutrition manufactures formula for wildlife. Amazon and Henry's Pets are good places to order. This formula composite is made to meet the needs of growing rabbits. There are two types. 32/40 is for under 4 weeks and the 20/50 is for babies over 4 weeks old.

Kitten Milk Replacer (KMR) powder can be used **TEMPORARILY** as it does not have the complete nutrients needed for healthy growth. However, it is easy to find locally in a pinch. Order Fox Valley asap.

Wombaroo is another wildlife formula brand sold by Henry's Pets. I honestly have never used it but have heard from several rehabilitators that think it's fabulous. It is pricey in part because it comes from Australia.

Wild animals have higher metabolisms and so have different nutrient requirements than your puppy or kitten. Talk to your veterinarian or an experienced wildlife rehabber to decide on a formula. Don't use social media as an authority for information. I have personnally taken in several animals who have been raised on a homemade milk replacer and they are far under weight compared to same aged peers.

MASS MARKET FORMULAS TO AVOID

There are a number of formulas that have been produced for the puppy and kitten market that do not meet the needs of wildlife.

DO NOT USE Condensed milk, cow milk, goat milk, human baby formula, or pet formulas. They are made for other species and will not meet the nutritional needs of a rabbit. Nutritional-related problems such as bloat, originate with poor formula choices.

HOW MUCH FORMULA?

Every infant will vary but there is a basic rule of thumb. Weigh the baby using a gram scale. The stomach capacity for a newborn rabbit is generally about 8% of the body weight. A baby over 5 days old can consume 12% of their body weight in formula.

For example, if a 3 day old baby weighs 25 grams then multiply by .08 or 8% of that which is 2ml's. We use cc's or ml's to measure the fluid. After you mix the formula you just draw up the correct amount into a syringe that nicely has the levels marked on it.

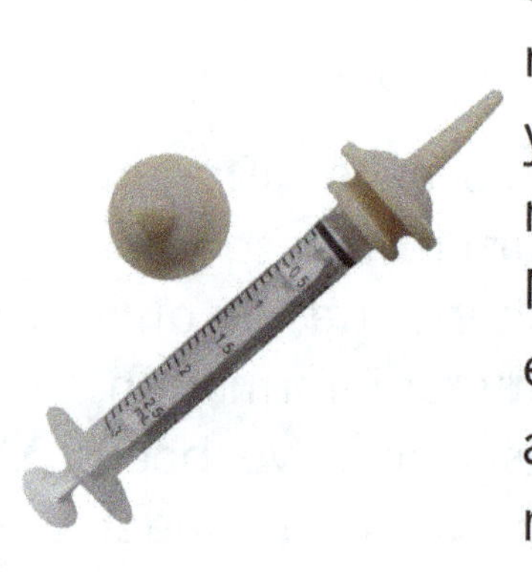

I use Miracle nipples. They were developed and patented by a squirrel wildlife rehabilitator! The "Mini" is a good size for young rabbits and then I bump up to the regular as they grow.

Many people like catac nipples. An eyedropper can work in an emergency but get a syringe and appropriate nipple asap. I use 1 ml syringes for newborns.

Stock up on nipples and small syringes. Nipples do wear out and they get chewed on by teething babies. Wash them gently in between feedings.

How To Mix, Heat, and Store

Mix powdered formula with very warm water. Follow the directions. Typically for Fox Valley it is one part dry formula to two parts water. Mix rabbit formula gently! Mixing vigorously can add air and cause gas or bloat. I recommend mixing a days worth of formula and then let it rest in the refrigerator. This allows it to dissolve thoroughly and for the air bubbles to rise to the surface.

Never heat the formula in a microwave. It doesn't heat evenly and will have hot spots that burn the baby.

Microwave a mug or bowl of just water. Heat to nice and warm but not hot. A good temperature is 105° F (40.5° C). Draw up the formula in the syringe and then place it in warm water. Also, baby bottle warmers can be useful.

Test it on your wrist before feeding (should be nice and warm). If the syringe formula cools during feeding then put it in the bowl of water to warm again.

Make formula in small batches and store in the fridge in between feedings. Discard any made-up formula after 24 hours.

WHAT ARE CC'S AND ML'S?

CC stands for cubic centimeter. ML stands for millimeter. They are just two ways that are used to measure volume in the metric system.

CC and ML are interchangeable. When you look at your syringe you want to pay attention to the number – not whether your unit says cc or ml.

Syringes can be purchased online or at local farm stores. For rabbits get small ones that are one cc or 1 ml. Make sure that the fractions are marked.

Use a kitchen gram scale to weigh each baby. Then multiply that number by 5% (.05) and that will be the number of cc's (or ml's) to feed per feeding. You may need to start with smaller feedings at first, especially on thin babies, and work up to 8%.

As They Grow Ask Yourself:
- Are they gaining weight steadily? Weigh them every day when they are small and record the data. To minimize handeling when you do their morning feeding, stimulate the bladder, weigh them, and change their paper towels.
- Are they active and exploring at an age-appropriate level?
- Do they appear sick or weak?

Older bunnies will be more comfortable laying down on a safe surface to feed.

Place animal in lateral recumbency (belly down). Place tip of nipple in mouth and give one drop; feed calculated amount slowly monitoring for aspiration (formula in the lungs; sneezing/coughing, formula coming out the nose). Some infants will happily (and quickly) suckle on a nipple, others will struggle and get stressed. Be patient and try various syringe tips/techniques for each individual as they will determine what they prefer.

They may scoot backwards. They know it's not mom! Place your hand behind their butt and gently hold them in place. . I will squeeze out a drop and hold it near their nose. Even though they know its a different smell they often recognize that it is food.

Never use pet nursers or doll bottles. The fluids flow to fast and can aspirate the baby. Aspiration is when fluids or formula enters the lungs. This can cause infection.
Don't turn the bottle so that it is facing down. This causes the formula to run faster into their mouth and the baby can choke. Feed in a quiet environment.

If fluids comes out of babies mouth or nose, then stop. You are going too fast. Gently turn the baby's head down while supporting body and allow fluid to come out. Then wipe his nose and mouth with a tissue. A human infant aspirator bulb can also be used.
Start over, slower. Allow them time to swallow. Babies can aspirate causing pneumonia, which can be fatal.
Use a small syringe not more than 3cc for neonates so that it releases a small amount. Baby bunnies have a small esophagus!

Infants need to be stimulated to urinate before each feeding. Gently rub the genital area with a cotton ball, Kleenex, or Q-Tip. First wet with warm water. Continue until cotton ball or other is no longer soiled.

You must stimulate cottontails until their eyes are completely open. Observe them to make sure they are urinating on their own. Use white paper towels or pee pads as cage substrate to visually confirm independent urination. Remember if you have a liter you want to determine that each one is peeing.

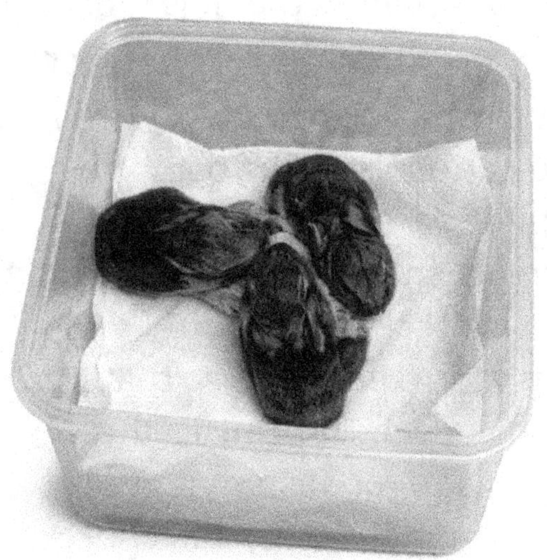

I like raising kits in a plastic tube because its easy to clean. I do use a lid (with air holes) because even infants can and do jump.

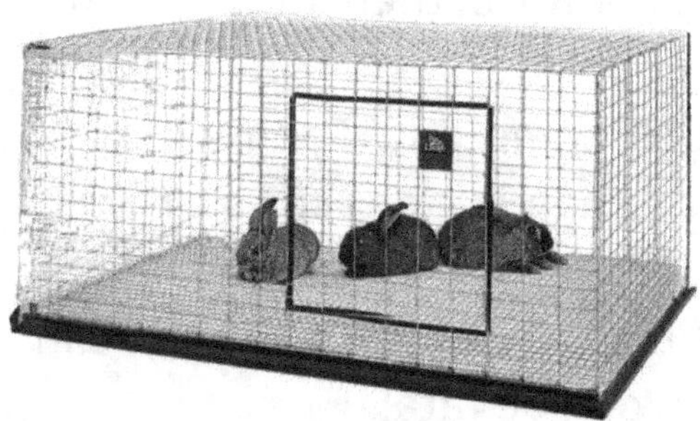

Good for young hares and older cottontails are the wire cages made for domestic rabbits. Make sure you get the small wire size for infant rabbits, often referred to as baby guards.

42

Rabbit Anxiety

Rabbits are prey animals. Their biology makes them vigilant for threat signals, and in unfamiliar or restrictive conditions they can become stressed or anxious. When we talk about anxiety in wild rabbits, we are not talking about mild nervousness. We are talking about a prey animal whose entire nervous system is built for survival. In wildlife rehabilitation, stress is not just behavioral. It directly affects their digestion, immune function, and recovery.

Wild hares and rabbits have a powerful flight response. When we have them contained or are holding them it causes stress because they can not naturally flee.

Cottontails rely on freezing and concealment. When stressed, they often:
- Press flat against a surface
- Close eyes tightly
- Become rigid
- Stop moving

Jackrabbits and hares are built for speed and distance. They stress rapidly in confinement and are extremely prone to capture myopathy. In rehab, jackrabbits require more space and more visual cover than cottontails. Hares may fling themselves against the sides of their enclosure.

Minimize Handling

Best practice:
Weigh quickly and efficiently
Perform procedures in one session when possible
Use towel restraint instead of direct gripping
Keep sessions short
Wild cottontails do not benefit from "bonding." Handling increases stress.
If they are struggling and have a rapid heart rate, place them in the enclosure (dark, warm place) and try agin in 30 minutes.

Provide Visual Security

Prey animals calm when they can hide.
Provide:

- Opaque nest boxes
- Deep grass hay for concealment
- Visual barriers between enclosures
- Covered sides on cages

Control Noise and Vibration

Rabbits detect low-frequency vibration and high-frequency sound. Housing should be:

- Away from barking dogs
- Away from predator species
- Free from loud music or sudden noise
- In low-traffic areas

This is fine for a domestic rabbit but don't let children handle wild rabbits. They should not be treated like pets.

Weaning

Why Weaning Is a High-Risk Period

The weaning period is risky because the young rabbit is switching from one type of digestion to another. If the timing is off, or if stress, poor diet, or pathogens interfere, disease can develop.

Most intestinal diseases in rabbit colonies occur just after weaning. These include:

- Coliform bacterial infections
- Coccidiosis
- Mucoid enteropathy
- Rotavirus-associated diarrhea

Weaning Foods

Dandelions, clover and plantain can be gathered from yards that have not been sprayed with pesticides or chemical fertilizers. Other greens that you can forage for are crabgrass, timothy, and bluegrass.

Note: Dandelions turn the bunnies pee yellow. This is natural, don't panic.

Organic romaine lettuce, endive, and mustard greens can also be fed.

Get good quality timothy hay from a pet or farm store. Even baby rabbits teeth grow continuously so offer small twigs for them to chew on. Dogwoods, mulberry, and apple are good.

Spinach, parsley, and beet greens should be limited. They contain oxalates which can bind calcium and cause a calcium deficiency.

Do Not Feed

Commercial pet rabbit mixes

Rabbit treats

Raisins, grapes, avocado

Bread, crackers, chips, cereal

Candy or sugary snacks

Supplements

Ame's soapbox!! I do believe that supplements can play an important role in the health of your rabbit. And as an organic farmer and herbal practitioner I do believe in herbal medicine. BUT there is definitely a time and a place. In addition, I believe each situation is unique. You need to consider the rabbits health needs, consult with your veterinarian and make an informed decision. That said here are some supplements I have found beneficial.

Valerian

Peggy Hentz, the founder of Red Creek Wildlife Center, recommends using valerian root. She uses the NOW brand valerian root extract and gives two drops to each bunny. It can be given by mouth or mixed into fluids or formula. She states that "Dosing is "per bunny" and not dependent on weight or age". (see resources for a link to her article)

Valerian is considered a medicinal herb used for insomnia, anxiety, and stress. For humans it is considered safe for short-term use. If you are a gardener, Valerian (Valeriana officinalis) is a hardy, fragrant, perennial that thrives in Zones 3-9. It is easy to grow in full sun to partial shade and likes moist, well-drained, fertile soil.

Probiotics

Bene-Bac Plus Pet Gel Probiotics is a product I have had good success with. It is helpful for preventing diarrhea and bloat. Contains live, naturally occurring digestive microorganisms.
It's available from Amazon and Henrys and many local pet stores. Probiotics are beneficial with rabbits who are taking antibiotics, recovering from illness or injury, or under stress. Add in probiotics to formula right before feeding.
Fox Valley also makes a probiotic called LA Probiotics.

Digestive Issues

Wild rabbits and hare kits have delicate digestive systems. Their gut flora (bacteria that live in the cecum) is immature at young ages, making them vulnerable to upsets when diet changes, stress occurs, or pathogens enter the gut. Wild kits get gut bacteria initially from there mothers milk and later from ingesting cecotropes.

Kits raised without a mother often do not develop a balanced microflora. This imbalance increases the risk of severe gut problems including diarrhea and bloat when solid food is introduced too early or formula is wrong.

Digestive issues are often a cause of death.

Steps For Prevention

- The first step is to make sure you have a clean environment so not to introduce bad bacteria. Washing your hands, surfaces, and disinfecting all materials used for feeding are critical.

- Secondly keep kits in a plastic tub with paper towels for bedding. The paper towels act as an early alert system. You can view the paper towels for urine and poop color and texture depending on the age of the baby. Normal poop is dark brown pellets. Light colored pellets that are grey or yellowish should be seen as a warning sign. See page 32

- Thirdly, digestive problems escalate quickly. Monitor but don't disturb. Stress makes the problem worse.

To complicate things, digestive issues in kits often overlap.
GI stasis stops gut movement → gas builds = bloat
Slowed gut allows harmful bacterial overgrowth = diarrhea

Signs of Digestive Problems:

Visual cues: Poop on their bottom, paper towels smears from runny poo, bad odor.

Smell: Kits do not have much odor (like fawns it's natures protection) so if you smell a sour ordor or a faint odor like sewer gas it is a immediate concern.

Hard belly: If you pick up a baby and find it's gut area feels tight or hard OR it may feel spongy. Learn what a "normal" belly feels like at each stage of development,.

Invest in A Microscope!

A microscope is a great investment. I discussed this in my first book, *Getting Started in Wildlife Rehabilitation*. Learning to do fecal exams will help you diagnose digestive problems, identify parasites, and spot some bacteria. Your vet can help you get started and there are several guides available as well.

Bloat

Bloat is a dangerous buildup of gas in the stomach or intestines. Rabbits cannot vomit , so if gas accumulates it quickly stretches the gut and presses on organs causing pain.

Bloat often develops suddenly and makes the abdomen firm, painful, and distended.

Even though it appears the baby is retaining water the first course of action is to give fluids. Sulfa drugs (Sulfamethoxazole/trimethoprim) treat bacterial buildup. They are used on humans for UTI infections and bronchitis.. Talk to your vet before baby season and have a plan! Having drugs on hand will allow you to move quickly for treatment.

Diarrhea and Enterotoxemia

Diarrhea in cottontail kits is often linked to an imbalance of gut bacteria (dysbiosis) or infection. When harmful bacteria overgrow, they can release toxins that damage the gut lining. Diarrhea can also be a symptom of internal parasites. So put the bowel movement under a scope and do a fecal exam or take a sample to your vet to check for parasite eggs.

Common contributors include:
Feeding formulas that are not species-appropriate
Early introduction of solid foods before the cecal flora is established
Contamination of feeding equipment or food
Stress and dehydration
Kits in rehab without appropriate exposure to maternal gut bacteria and balanced diet are particularly at risk.

Treatment includes rehydration. Skip the next formula feeding and replace with fluids.
You can treat with Kaolin Pectin, a veterinary OTC drug made for livestock and cats and dogs. Get one labeled for animals such as Durvet. Do not use the human Kaopectate or Pepto Bismal drugs as they are not good for rabbits.

Gastrointestinal (GI) Stasis

GI stasis is one of the most serious and common life-threatening digestive issues in rabbits. It occurs when the normal movement of the gut slows down or stops entirely. Rabbits normally need to eat and pass feces continuously. If gut movement slows:

- Food and gas build up
- Pain occurs
- Gas accumulates (bloat)
- Beneficial bacteria decline and harmful bacteria can overgrow

Coccidia

Coccidia is a protozoa that lives in the digestive tract and is common in rabbits. They are actually present in the digestive system of healthy animals and don't necessarily cause a problem. Stress can cause the protozoa to multiply which typically presents with diarrhea. This can lead to dehydration and death.

Coccidia is very contagious and is transmitted via ingestion of contaminated food or feces. The eggs called oocysts can be seen under a microscope.

Toltrazuril 2.5% can be used to treat. Make sure to follow directions as it is often given on an irregular schedule. The Merck Veterinary Manual suggests using Sulfaquinoxaline to help control it.

Disinfection and keeping a clean environment is important. Ammonia (10% solution) will kill oocysts and is useful to disinfect cages and surfaces. Have an extra enclosure to put kits in and clean with ammonia outside as the fumes are not safe for baby animals.

Drugs To Avoid

Rabbits have sensative systems and react badly (or die) if given certain medications. Talk to your vet about medicines.

Never give rabbits any drug ending in cillin. **No penicillian or amoxycillian.**

No enrofloxacin an antibiotic commonly called Baytril

Never use **fipronil (Frontline)** as it is toxic and can be fatal to rabbits.

Rabbit Pests and Diseases

Rabbits get a number of pests and diseases. Some are easy to manage and some take more time (and money). It's important to give your new intake(s) a thorough exam to look for symptoms and signs of problems.

Zoonotic diseases or zoonoses are diseases that can be transferred from animal to human or from human to animal. As wildlife rehabilitators, we need to be aware of them as many are potentially dangerous.

Rabies is probably the most well known zoonotic disease. Fortunately, rabbits rarely get rabies. The Wisconsin Department of Health states that "Small rodents (e.g., squirrels, hamsters, guinea pigs, gerbils, chipmunks, rats, and mice) and lagomorphs (rabbits and hares), whether wild or kept as pets, are rarely found to be infected with rabies and have not been known to transmit rabies to humans". I do not vaccinate rabbits for rabies. However, I do recommend all wildlife rehabilitators get the Rabies Pre-exposure Prophylaxis shots.

Rabbits can carry the zoonotic diseases tularemia, plague, ringworm, and salmonella. Some basic common sense things are to wash your hands before and after handling wild animals, wear protective clothing, and dont eat in the room where you handle wildlife especially new intakes.

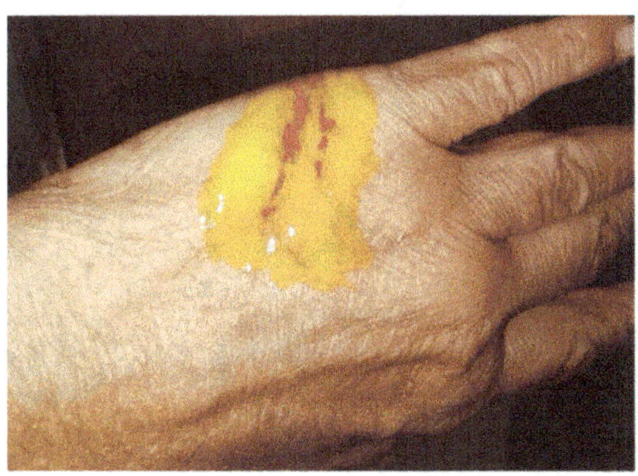

All animals can and do bite. Here's my hand after a squirrel attack

External Parasites

 External parasites live on the outside of the animal's body. You may see their eggs, fecal matter, or bodies during an exam. These external parasites are often seen on baby wildlife.

The good thing about external pests is that you can see and identify the adult quite easily. Nymphs and larval stages are small and you may need a magnifying lens. Treatments are often easy to apply and purchase.

The challenge with external pests is that they are prolific, and it may take repeated treatments to get them under control. Ticks and fleas have both been found in Antarctica where they feed on cold weather birds such as penguins.

Also, if you are not careful and diligent, they can easily spread throughout your home. Have an examination area such as a countertop or stainless steel table. An area you can disinfect and clean easily. Don't pick up a new arrival and hold it against your chest. That is just inviting the parasites to climb on you.

Orphaned baby wildlife is prone to parasites because they have not had a mother to groom them. A heavy load of external parasites may result in anemia, weakness, and secondary skin infections. So treating them ASAP is important.

Ticks

Eastern cottontails are hosts for many tick species . Ticks may attach to the rabbit's ears, face, and neck, where the skin is softer. They can carry a wide range of ticks including the rabbit tick, Asian longhorned tick, Lone star tick, and the American dog tick.

Rabbits are sensitive to chemicals. Pyrethrin can be used for tick and flea control. Make sure it is labeled for kittens. Zodiac Flea & Tick Powder is one that I have used. Use caution and don't just shake it on an infant. Use your hands or a cotton ball or Q-Tip to spread it carefully.

The Rabbit Tick

The rabbit tick (Haemaphysalis leporispalustris) is wide spread throughout the Americas. Rabbit ticks are a brown tick with a triangle shaped mouth. They are common on both cottontails and hares as well as ground feeding birds such as grouse.

The rabbit tick is not considered a vector in transmitting Lyme disease to humans however caution and care is important when handling infested animals. They can carry Francisella tularensis (tularemia). In addition, they can get on pets.

Photo credit:
Cricket Raspet

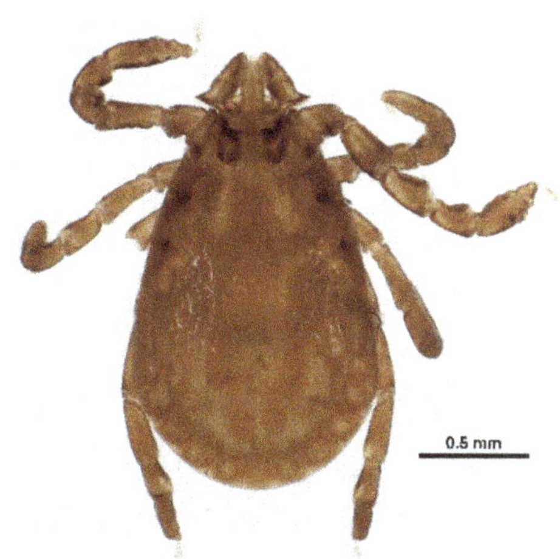

Photo credit:
Centre for Biodiversity Genomics

Removing External Parasites

I am all in favor of treating pests naturally as I was a certified organic farmer for many years. However, often in wildlife rehabilitation we need to act quickly to save the animals life.

For compromised rabbits, I start with a chemical treatment and then use a natural treatment to control the pests. Natural products are great, but they don't often kill the parasite quickly thus eliminating the issue. In some cases, time is essential to save the life of the animal.

Non-chemical ways
• Keep incoming rabbits in a quarantine area away from other animals to avoid transmission.
• Apply insect repellent to your clothes so they don't get on your body and use you as a transportation mode.
• Ticks and maggots may be easy to pick off by hand. Use tweezers or a special tick-removal tool.
• The CDC recommends using tweezers to grab the tick as close to the skin as possible. Pull upwards without twisting. Don't squish the tick because that can spread disease through body fluids.
• Flea and lice combs are also useful to remove them and then dunk the pests in soapy water. The risk here is causing stress.
• Fleas can sometimes be removed by using a lint roller or masking tape to catch them. This can work well in an infant that hasn't grown fur yet.

  • Products that are made for domestic kittens can generally be used for rabbits. Caution is needed to use the correct dosage for the type of treatment.
  • Use products with plant-based and natural ingredients. Vet's Best has a line of natural flea and tick repellants. Keep in mind that natural ingredients can still be strong so look for products ok for kittens or pet rats.

Maggots

Maggots are fly eggs. The fly's lay the eggs on exposed flesh such as a wound. When the eggs hatch the larvae (worms) will eat the rabbits flesh. Clean the wound with saline solution (I will cover more in the wound section) and pick out the fly eggs and larvae.

Capstar® (Nitenpyram) is an over the counter treatment that kills maggots as well as fleas. It is easy to find on Amazon or Henry's Pets. Capstar comes in two sizes. Get the 11.4 mg For young rabbits I prefer to grind up 1 tablet and mix it with 29 mL of warm water. After it dissolves spray it on the rabbit avoiding the eyes and mouth. You can use a cotton ball soaked in it for young kits.

Maggots (fly eggs)

Examples of mange on a deer and a fox

Mange/Mites

There are several types of mites. Mites are common in ears and under the skin. They are ridiculously small, and some are microscopic.

Mites are arachnids that burrow under the skin. They reproduce rapidly and can also infect bedding and cages.

MANGE

The most severe mites that I see are mange. It is more likely you will see them on juvenile or adult animals. Mange is more commonly seen in larger mammals like foxes and coyotes.

Some rehabilitators use medications like ivermectin or selamectin to treat the mites. Ivermectin is a commonly used antiparasitic for treating ear mites, fur mites, and sarcoptic mange in rabbits, however it is a strong drug. Generally considered safe, improper amounts can cause toxicity so talk your vet about doseage.

Mange has several symptoms:
- Severe itching
- Skin that is scaly, and irritated looking
- Distinct bad smell from dead skin and infection
- Hair loss especially face, legs and groin
- Severe causes leave crusty lesions on the body
- Swelling especially around eyes

Mange is serious. You can check for mites by viewing a skin scraping under a 4X setting on a microscope. Typically the type of mange to effect rabbits is Sarcoptes scabiei (Sarcoptic mange).

The mites typically start around the head and can spread to the entire body in a matter of weeks.

The entire mite life cycle is spent on the host and lasts about 21 days. It is transmitted by direct contact and the young often get it from the parent.

CLEAN AND ISOLATE

Mange is very contagious and will spread to other animals in your care. In addition, humans can get mites. They can not reproduce on your body, but they can dig in and chew on your skin. This will cause intense itching and pain.

• Keep the animal with mange in isolation.
• Wear gloves when you treat, feed, or clean the mange animal. Wear protective clothing and insect repellent because while they don't live on humans they will bite you.
• Bedding should be changed daily, placed in a sealed garbage bag, and thrown out.

Ear Mites

Ear Mites (Psoroptes cuniculi) are non-burrowing and cause severe brown crusty inflammation in the outer ear. Symptoms include crusty layer plus intense itching, foul odor, and head shaking. There is potential for secondary bacterial infections.

Don't scrub the infected skin because it will cause intense pain. The tissue will heal itself as the mites die during treatment. Ivermectin and moxidectin are commonly used treatments.

It is not zoonotic but will spread between rabbits.

Clean, healthy ears
are important
Antelope Jackrabbit

Fleas

Fleas are smaller than ticks and hop as their mode of locomotion. You may see pepper-like flecks on the skin which are flea feces. The larvae feed on these feces. They will infest bedding areas and can easily spread to other areas of your center.

Fleas and small mites are common when baby rabbits first come from the nest. I recommend using kitten flea powder, as it is usually safe for rabbits too. You can use gentle sprays for birds.

When treating, don't put the powder directly on the rabbit. Use a cotton ball or cloth to gently apply sprays or powders. Zodiac Flea & Tick Powder is one that I recommend as I stated in the tick section.

Important note: Beware of using harsh commercial products. Buy products from your vet, pet store, or a reliable brand. Many cheap products sold at big box stores are produced in China or India and contain strong chemicals that can kill baby animals.

juvenile
white-tailed
jackrabbit

Rabbit Bot Fly, Cuterebra buccata

The bot fly or warble fly, is a more unusual pest that affects rabbits and hares in Eastern North America. They are especially common in the southeast due to warmer winters.

The botfly lays its eggs on plants, and the rabbit consumes the eggs. Larvae can also crawl onto the rabbit's body from the environment. Inside their bodies, the eggs hatch into larvae and then move to areas under the skin. As they grow, you can see a lump forming, which soon develops a breathing hole. If left alone, the lump will become very large and oval-shaped. In late summer you can see the large swollen skin sacks that are caused by the growing larvae.

Use tweezers to remove the bots from the sack. (Gross I know!) Then clean the affected skin and apply first aid cream. Not many studies have been done on effective treatments but talk to your vet if they look severe.

If you have read my Squirrel Rehabilitation Book you know that bots are a big problem in squirrels as well. This Eastern gray squirrel has swollen skin and wounds caused by the tree squirrel bot fly.

Internal Parasites

There are several types of worms that infect rabbits.

Nematodes are the most common rabbit worms and are commonly called pinworms. They get pinworms by ingesting the eggs in their environment. They are zoonotic.

Symptoms include diarrhea, weight loss and lack of energy. You can see the worms under a microscope. Worming medicines are available and some work better than others on certain worm species. Also, some wormers are quite strong, and care should be taken with young, compromised bodies.

Worming doses are very small so this is the time to use those 1ml syringes. That way you can measure accurately. Check with your vet for their recommendations.

Panacur (fenbendazole) can be used for rabbits. You can often find the liquid in farm stores. For the 100mg strength dose is typically 0.05mg per 100 grams of weight. MED Animal Health makes Panacur® Rabbit 18.75 % Oral Paste - obviously geared towards domestic rabbits. Consult your vet for dosage.

Trematodes and Cestodes, commonly called **flatworms**, common in juveniles and adults. May crawl out of the anus and be visible. Tapeworms in rabbits are primarily transmitted by ingesting vegetation, hay, or water contaminated with the feces of definitive hosts. They are carried by fleas. Use Praziquantel wormer to control them.

Wounds

The first step is stabilizing the rabbit. As long as it's not an emergency treat for shock and place in a warm dark place before other treatments are attempted. Wounds need to be cleaned, treated and protect skin so it can heal. Depending on the wound you may need to control bleeding.

Do not use hydrogen peroxide. New research states that it can damage skin cells and healthy tissue. I use chlorohexidine. A solution of 1 % Chlorhexidine or Betadine. I draw it up in a syringe and flush the wounds and apply a dressing as needed. Some brands may need dilution. Deep gashes may require stitches. Another option for cleaning is a saline solution.
Bandage wound so the rabbit can't lick medicine and to protect from dirt/feces. A bandage also helps the area stay moist.

Abscesses

Abscesses are often caused by a cat bite or other animal attacking the rabbit. They look like a swelling or lump and are full of pus. The mouth of the attacking animal contains bacteria. You should consult with your veterinarian to discuss treatment. Abscesses often require antibiotics. They also may need to be punctured, drained, and cleaned. Applying a warm compress can help it drain.

Antibiotics can often cause digestive problems and diarrhea so extra fluids and probiotics are helpful.

Debridement

Debridement is the removal of damaged tissue or foreign objects from a wound. Dead or necrotic tissue is often present in an older wound. When you remove the dead tissue it allows new healthy tissue to grow again. You can use small surgical scissors and surgical scrub to remove dead tissue if the animal will tolerate it.

Degloving

Degloving is when the skin and often deeper tissues are separated from the body. These incidents can occur when the victim is attacked by another animal or in a mechanical injury. The skin is basically ripped off the body. There skin is designed to tear away as a defense mechanism to help them escape predators. You may see exposed raw flesh/muscle, hanging flaps of skin, and maggots. You may see this more in juveniles and adults than in infants.

The degloved animal is often in shock. They are in intense pain and may scream. Any bleeding should be controlled first and then the wound needs to be cleaned with a sterile saline solution. One option is to apply topical silver sulfadiazine (also great for burns). Silver sulfadiazine is a prescription drug. Another option is Neosporin but it may not have the strength needed for a more severe case.

Bandage and wrap the area. Give Meloxicam (Metacam prescription) for pain and inflammation. 0.3- 1.0mg/kg twice a day.

Gunshot Wounds

Sadly, wildlife rehabilitators do receive animals that are shot in an effort to get rid of a nuisance animal or used for target practice. Bullets and BB's push hair and dirt into the body often leading to infections. Depending on the bullet the rabbit may have bone, muscle, and/or vascular damage. They should be treated as open wounds. Don't close the wound immediately since the risk of contamination is so high.

If a shot animal is 'out of season' or on the endangered species list take photos, save the bullet fragments, and immediately report it to your local Fish and Wildlife officer.

Broken Bones

Rabbits may have broken legs when admitted. It may be obvious by angle of leg or you may observe swelling. Typically neonates bones will realign and heal in the correct position.
Metacam (meloxicam) is a great drug and anti-inflammatory that reduces swelling.

You may suspect a broken back. This may occur when a rabbit or hare is trying to escape from danger or a baby who is trying to jump out of a container. You will be surprised how high a young rabbit can jump!

Spinal injuries are a wait and see type and I do collaborate with my vet because I don't want to prolong agony for them. But giving them time to heal is important. That and a warm, quiet, dark place.

Adults that come into your care may need to be given fluids like any other intake. In addition, they may be in shock and feel chilled. So place a heating pad partially under the cage. This can be tricky because with a spinal injury they may not be mobile. They may not have much interest in food but often I offer half strength formula in a syringe. I want them to get some nutrients but still go easy on there system. After a day or so as they feel better Ill offer some soft foods such as baked sweet potatoes and cooked broccoli.

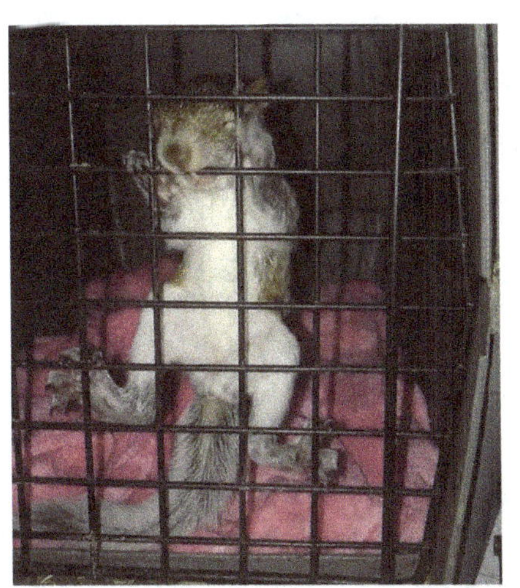

If they come back from the
vet looking like this, they
feel better!!!
Photo credit: Ame Vanorio

Diseases

I will list some of the more common illnesses that you may encounter while rescuing rabbits and hares. Work with your veterinarian to establish treatment plans.

West Nile Virus

West Nile Virus (WNV) is primarily known for affecting birds, humans, and horses, but it can also infect other animals, including rabbits. Like squirrels and chipmunks they are considered a dead end host but can act as a reservoir in nature.

The primary mode of transmission for WNV is through the bite of an infected mosquito. Infected rabbits don't always show symptoms. Some might show signs of neurological issues such as tremors, lack of coordination, difficulty moving, and weakness.

Supportive care is needed. Hydration and tube feeding may be necessary if theyl are having a hard time swallowing. NSAIDs such as Metacam are helpful. Create a quiet, stress-free environment. A warm, comfortable, and safe enclosure is essential for recovery.

There is no cure but most animals make a full recovery. Some may display permanent neurological symptoms.
West Nile is transmitted by mosquitos so reducing their numbers around your home or center is smart. Change outside water dishes frequently and don't allow standing water to accumulate. Mosquitoes lay their eggs in moist soil and water and they typically hatch in 48 hours.

Rabbit Hemorrhagic Disease

Rabbit Hemorrhagic Disease (RHD) is a highly contagious viral disease affecting that affects rabbits, caused by the RHDV2 virus. It is spread by direct contact.

It causes severe hepatitis (liver damage) and rapid death, often within 48 hours of infection. RHDV viruses are highly contagious with high mortality rates. There is no cure. Clinical symptoms may be absent but may also include vocalizations, bleeding from the mouth, nose or rectum, fever, and difficulty breathing.

Cornell University states that the New England cottontail (Sylvilagus transitionalis) and the snowshoe hare (Lepus americanus) are likely to be susceptible. RHDV2 does not infect humans. However humans can carry it on clothing or dishes.

As of this writing in early 2026, Rabbit Hemorrhagic Disease Virus 2 (RHDV2) is considered stable-endemic or has been detected in wild and domestic rabbit populations in 14+ states. Wild rabbits are an important food source for carnivores and raptors, RHD is a concern because it also disrupts the food web and can harm these animals as well.

For wildlife rehabilitators RHDV2 is a reportable disease. Euthanasia is often the humane option. I discuss euthanasia more in my "**Getting Started in Wildlife Rehabilitation**" book.

Tularemia ("Rabbit Fever")

Tularemia ("Rabbit Fever") is a bacterial disease caused by *Francisella tularensis* a gram negative bacteria. It often kills rabbits and hares but it can also affect opossums, skunks, and squirrels. It has a short incubation period and can infect within 4 to 10 days. The disease progresses quickly and animals may die without having symptoms. Infected animals may become lethargic, weak, and uncoordinated, frequently showing skin ulcers, and abscesses.

While it is rare it should be taken seriously as the disease is often fatal, causing local population crashes. It is spread via ticks, deer flies, or direct contact with infected animals.

It is **Zoonotic** and can infect humans through bites, ticks, or contact with carcasses. It causes swelling in your lymph nodes and Tularemia should be treated as soon as possible with antibiotics. It can also infect domestic animals.

Tularemia is considered a reportable animal disease in the United States. American Veterinary Medical Foundation (AVMF) states It has been found in every state except Hawaii.

The organism can survive for weeks to months in a moist environment. Clean thoroughly with disinfectants. Wear gloves when cleaning enclosures.

Metabolic Bone Disease

Metabolic Bone Disease (MBD) is due to inadequate nutrition and can develop in rehabilitation. It is more common in squirrels but can also affect rabbits and opossums. MBD is a condition resulting from an imbalance of calcium, phosphorus, and vitamin D3 in their diet.

This disease is totally preventable and curable in most cases but sometimes has permanent bone curvature requiring a forever home.

Causes:
- Calcium Deficiency: Lack of sufficient calcium in the diet.
- Phosphorus Imbalance: An excess of phosphorus relative to calcium can disrupt the balance necessary for healthy bone development and cause blood calcium levels to fall.
- Vitamin D3 Deficiency: Vitamin D3 is crucial for calcium absorption in the body. Lack of sunlight or dietary sources of vitamin D3 can lead to deficiencies.
- Starvation
- Weaning to early.

Symptoms:
- Weakness and Lethargy:
- Bone Deformities: Softening of bones (osteomalacia) can lead to deformities, fractures, and abnormal bone growth.
- Lameness and Difficulty Moving: Drags hind legs.
- Tremors and Muscle Twitching: Neurological symptoms such as tremors and muscle twitching can occur due to calcium imbalances affecting nerve function.
- Swollen Joints: Visible swelling in joints and limbs.
- Poor Appetite: A decrease in appetite and overall condition resulting in loss of weight.

MBD Con't.

Treatment:

- Dietary Correction: Ensure a balanced diet rich in calcium and appropriate levels of phosphorus.
- Give plenty of timothy hay and alfalfa hay in small doses.
- Calcium Supplements: In severe cases, calcium supplements may be necessary under veterinary guidance.
- Vitamin D3 Supplementation: Providing adequate vitamin D3, either through diet, exposure to natural sunlight, or supplements.
- Exercise is important for bone development.
- Supportive Care: Providing a safe and comfortable environment. They may appreciate a heating pad in one corner. Pain management.

Snowshoe Hare

Shope papilloma virus (SPV) or cottontail rabbit papilloma virus (CRPV)

Shope papilloma virus (SPV) is a naturally occurring virus that infects wild rabbits and hares, especially the eastern cottontail. The virus infects skin cells and cause benign tumors called papillomas or warts.

These raised, horn-like growths on the body. spread mainly through vectors such as mosquitoes and biting insects. The virus enters through small breaks in the skin created during insect feeding. The most obvious sign of infection is the development of large wart-like growths on the skin.

The virus is endemic to the high plains/Midwest regions. Typically young rabbits die from the disease. Adults may fight off virus. Prevention is treating all incoming animals for parasites and controlling biting insects.

Rabbit with Shopes Papillomavirus affecting the mouth
Photo credit: WD45

Teeth Issues

Rabbits and hares have teeth that are open-rooted, and grow continuously. Dental issues, such as malocclusion (misalignment), broken teeth, or abscesses, occur when teeth do not wear down properly, leading to severe pain, inability to eat, and potential starvation. You may see this in juveniles or adults that come in. Severe dental distress will need a veterinarian who may need to clip, remove, or file teeth.

Common Teeth Issues

Malocclusion: Misaligned, overgrown, or broken teeth. It typically affects the front incisors and they may grow in abnormal angles.

Overgrown Molars: Rabbits back teeth can develop sharp, spurs that dig into the tongue or cheeks. This can cause severe pain and reduced appetite.

Trauma: Broken or fractured teeth from accidents, which can lead to uneven growth or abscesses.

Symptoms of Dental Pain in Rabbits

- Weight loss and poor body condition.
- Reduced appetite or picking at food (anorexia).
- Drooling (hypersalivation) or a wet chin.
- Weepy eyes or a runny nose caused by root pressure on tear ducts.
- Bumpy/swollen jaw.

Examining teeth on my domestic rabbit

Housing

House rabbits with others that are the same age and size. All rabbit housing should be away from humans and pets in a private location. Clean cages once daily for eyes-closed babies. Older rabbits may need their cage cleaned twice daily as they will hop all over and will poop on food.

Infant rabbits with their eyes closed can be kept in plastic tubs, plastic critter carriers or aquariums. Paper towels, pee pads, or a soft towel serves as a good liner for infants. Ensure substrate is in good condition as frayed edges/strings can be dangerous.

Always provide a hiding house for eyes open rabbits to help prevent stress. Hiding places can be a box for them to crawl into, or anything they can huddle inside. The half "logs" sold for reptiles work well when they are small. If you use a transparent container such as an aquarium place a towel over one half to help them feel safe.

Having a clean rehab area helps keep animals and people healthy. Wear protective clothing such as gloves and a face mask especially if you suspect or know about an illness. Have another space to put babies while you are cleaning so they don't inhale any cleaners. This sounds weird but spraying disinfectant on dirty bedding will help reduce dust and germs from going into the air. Once the bedding is moist remove it, wipe down the bottom of the enclosure, let it air and then move the babies back in.

Outdoor housing should be completely protected from predators including domestic pets. They should be in a safe location and be sturdy. Double walled cages work well You can reinforce them with ½" hardware cloth and locking latches to make them predator proof.

IWRC and NWRA recommend in their book **Minimum Standards For Wildlife Rehabilitation** (see resources) that when designing housing for rabbits and hares avoid using wood, chain link, wire mesh, or hardware cloth as the sole materials in construction. Rabbits do not have good depth perception and will not "see" the fencing. They recommend using "Sight barriers" at the height of the adult animal's ears (12"-24") made of shade cloth. For outdoor, above-ground enclosures, the bottom should be constructed of 1/4-inch mesh for drainage, and covered with hay to prevent foot trauma.

Eastern Cottontail babies can be kept in a tub or tank that is at least 10 Gallon or 38Liters. Juveniles and weaned babies need 6 ft x 6 ft x 4 ft or 1.8 mx 1.8 m x 1.2m. Increase size for litters of more than 4 kits.

Many people use commercial rabbit enclosures. Make sure the wire spacing is 1/2 inch by 1/2 inch on the floor and extending 6 inches up the sides. Weaned rabbits can use 1/2 inch by 1 inch. Look for Use 14-gauge or 16-gauge wire. It's cost effective to make your own. See page 42 for more examples

Obviously this is a chicken cage but something large and secure for hares during pre-release works well.

I used chainlink fence for fox pre-release with additional chainlink on top and bottom. In addition, the lower third was wrapped in smaller wire. I think if the entire structure had an additional layer of 1x1 wire it could protect hares.

Bleach

Sodium hypochlorite, commonly known as bleach is a good disinfectant. It is also not as harmful as some of the other chemicals used for cleaning (and its cheaper) 1 tablespoon of bleach per 1 quart of water is a good strength. Thoroughly disinfect tubs in between liters or if you have a contagious disease. Let the tub or aquarium sit outyside in the sun after disinfecting.

Jackrabbit Considerations

Jackrabbits have a long breeding season and kits can be born between February and September. There are usually 1 to 4 in a litter. Kits are born in a low spot or shallow depression scratched into the soil. They don't make a fur lined nest.

Jackrabbits are born with eyes and ears open, fully furred, and mobile. However, they are still dependent on nursing. They start foraging within two weeks but still nurse daily. Jackrabbits do not wean as quickly as cottontails and sometimes will want/need formula to 6 weeks. Its important to monitor their feeding behavior and body condition to determine when to release.

Minimum standards recommends 20Gallon or 76 Liters for newborn jackrabbits. For weaned babies an outdoor enclosure should be at least 10 ft x 10 ft x 4 ft or 3.0 m x3.0m x 1.2m. When they are @ 250 grams, let them out for nightly exercise. Have a larger cage or allow them a private room. They need to exercise.

If you have an outdoor pre-release are you can make it feel like a natural habitat. A leanto or triangular structure that has two openings works well for hiding spaces. This works well for hares especially for jack rabbits, as it has two exits so the animal doesn't feel trapped.

A carrier placed in the cage may act as a den and be used for capture as needed to move to release location. Small shrubs, potted plants, or bales of hay also can be used to provide shade and shelter for rabbits. Provide branches of edible species for gnawing to curb tooth growth should be readily available. Jackrabbits grow quite quickly and can be surprisingly strong.

Release

The purpose of wildlife rehabilitation is to be able to release the animals and let them experience a normal life. One of the great things about rabbits is that they are ready to release in a relatively short time. Releasing cottontails depends on a combination of behavior and numbers (weight and age).

Typically, the age to release is between four to six weeks with a healthy body condition. If you hold onto a rabbit too long, they can get stressed and die (captivity/capture myopathy).

All orphans should be acclimated to outdoor weather conditions. You can do this on a screened in porch, by opening the window in their rehab room, or having an outside enclosure for them.

Solid foods should be their only diet for a minimum of one week prior to release. Behavior to watch for includes boxing (jumping at your hand), trying to escape, digging, and if there is any aggression between juveniles, it is time to release your rabbits.

Release sites should be as safe as possible, an area away from predators (including cats), that has plenty of food sources and hiding areas. You will want to have many sites you can use. Release them into a brushy area adjacent to grassy areas with plenty of hiding places.

The best time to release them is near dawn. That gives them a chance to find an appropriate hiding places before nightfall.

Best practice release criteria:

- Correct weight and body condition
- Strong mobility and coordination
- Normal fear of humans
- Able to eat appropriate wild foods
- No signs of illness
- Weather appropriate (avoid extreme heat/cold, severe storms)

Best practice release sites:

- Where the rabbit was found when possible and legal, because local habitat is familiar and often best.
- Good cover and food (brush, edge habitat, native grasses/forbs).
- Low immediate danger (not next to outdoor cats, busy roads, or high predator concentration from feeders).

Male jackrabbits often have
dramatic displays during courting

Rabbits and Hares in North America

Here are 15+ rabbit and hare species native to North America, not including subspecies. I am going to look the more common ones we see in rehab. As a wildlife rehabilitator you need to learn which species are in your region, their behavior, breeding habits, and what they eat.

Eastern Cottontail: *Sylvilagus floridanus*

The eastern cottontail is the most common rabbit in eastern and central North America. They lives in fields, brushy edges, open woodlands, suburban yards, and farms.

Breeding starts in late winter and can continue into early fall. A female may have several litters each year. They generally remain within the same area throughout life. Home ranges average 1.4 acres.

This species is unique because it thrives in human-altered landscapes. It has expanded its original range from the Midwest, westward and eastward, sometimes replacing native cottontail species like the New England cottontail (Sylvilagus transitionalis).

Desert Cottontail: *Sylvilagus audubonii*

The Desert Cottontail aka Audubon's cottontail, is a rabbit species native to arid regions of the southwestern United States and northern Mexico. It is known for its adaptability to dry habitats. This species plays a key role in desert ecosystems as both herbivore and prey for many predators.

Desert cottontails give birth to their kits in burrows self-dug or abandoned dens of other animals. Breeding occurs nearly year-round in warm climates, with females producing multiple litters annually. Each litter contains two to six young which mature quickly. In the wild, most individuals live less than two years due to predation and environmental pressures.

Mostly crepuscular, Desert Cottontails feed during dawn and dusk to avoid midday heat. They rely on stillness and camouflage when threatened, resorting to rapid, zigzag running when chased. They can sprint 20 mph and climb sloping trees and thick bushes to reach food or escape danger. Their huge ears are used to radiate heat.

Photo credit: Jessie Eastland

Mountain Cottontail: *Sylvilagus nuttallii*

The Mountain Cottontail is a small rabbit species native to western North America. It inhabits dry shrublands, grasslands, and forest edges, particularly in mountainous and foothill regions.

The Mountain Cottontail is primarily crepuscular, it feeds on grasses, forbs, and woody vegetation. In winter, it shifts to twigs and bark when herbaceous plants are scarce.It prefers areas with dense cover, such as sagebrush or juniper stands, interspersed with open feeding zones. They do not hibernate and remain active all year, often hopping in deep snow to find food. Like some rabbits and hares they do not change color to camouflage.

Breeding occurs from early spring to late summer, with females producing two to five litters per year. Litters usually contain three to eight kits, which develop rapidly and disperse within a month. High predation rates contribute to a short life expectancy.

Mountain Cottontails are vital to regional ecosystems, serving as prey for hawks, coyotes, foxes, and bobcats. Their browsing affects vegetation dynamics, influencing plant community structure and seedling survival in arid and montane habitats. Like the desert cottontail they can climb bushes and low trees, such as junipers, to find food.

Photo credit:Vanessa Johnson

Photo credit: Justin Wilde

New England Cottontail: *Sylvilagus transitionalis*

The New England cottontail is native to the northeastern United States. It depends on dense young forests and shrub thickets. Breeding occurs from spring through summer. It nests in thick cover to avoid predators.

This species is unique because it is in decline due to habitat loss. Once distributed widely across southern New England and eastern New York, the species now occupies fragmented patches. It has habitat pressure from the more adaptable eastern cottontail.

Photo credit:Jessie Brown
New England: Roger Williams Park Zoo

Appalachian Cottontail
Photo credit: Kristof Zyskowski

Appalachian Cottontail: *Sylvilagus obscurus*

The Appalachian cottontail lives in high elevations of the Appalachian Mountains. It prefers spruce-fir forests and dense mountain shrublands. Thick understory vegetation is important for cover.
Breeding takes place during spring and summer. Females build shallow nests similar to other cottontails. Young develop quickly in the short mountain growing season.

This species was once confused with the New England cottontail. Genetic research confirmed it is a separate species. It is adapted to colder mountain climates.

80

Swamp Rabbit: *Sylvilagus aquaticus*

The swamp rabbit is a large rabbit found in the southeastern United States. It lives in wetlands, bottomland forests, and along rivers. It prefers thick vegetation near water. It often uses raised areas during flooding.

The species is notable for its semi-aquatic behavior, unusual among North American rabbits. It is adapted to wetland habitats and is a strong swimmer. They often use water to escape from predators.

They have short, rounded ears, coarse brownish-gray fur, and a reddish nape. Their hind legs are strong, aiding both in rapid hopping and swimming. The fur on their feet helps them move efficiently through wet environments.

Breeding begins in late winter and can continue into summer. Nests are built on high ground in dense vegetation. Juveniles are weaned within a few weeks and reach maturity in several months.

Photo credit:Dan Vickers

Photo credit: Brandon Johnson

Marsh Rabbit: *Sylvilagus palustris*

The marsh rabbit lives in coastal marshes of the southeastern United States. It occupies both freshwater and salt marsh habitats. Females build nests in grassy clumps above water level. This species is smaller than the swamp rabbit. It is strongly associated with coastal habitats. It can swim well and uses water for escape.

Marsh rabbit. Photo credit: Richard Stovall

Juvenile Bachman's Rabbit

Bachman's Cottontail: *Sylvilagus bachmani*

Bachman's cottontail lives along the Pacific coast, especially California. It prefers chaparral, coastal scrub, and dense brush. It rarely ventures far from cover. Fire and vegetation patterns influence its habitat.

Breeding usually occurs from late winter through summer. Nests are hidden in thick shrubs. Young grow quickly and disperse within weeks.

It tolerates dry summers and mild winters. Habitat fragmentation can affect local populations.

Snowshoe hare: Lepus americanus

The snowshoe hare is native to the boreal and mountain forests of North America. Named for its oversized, fur-covered hind feet, it is renowned for its seasonal coat color change that provides camouflage in their environments. They go from brown or grayish in summer and pure white in winter. This change, triggered by day length rather than temperature, occurs over several weeks.

Primarily nocturnal and solitary, snowshoe hares feed on a variety of vegetation—grasses and herbs in summer, woody twigs and bark in winter. They are powerful jumpers and can leap 10 to 12 feet in a single bound to help escape from predators.

Breeding occurs from March through August. Females bear two to four litters per year, each with one to eight leverets born fully furred and mobile. The young mature quickly, reaching independence within a month and breeding age within a year. Most snowshoe hares live one to three years in the wild, though some survive up to five years.

Snowshoe hares do not hibernate. They stay in the same general area as where they were born, typically living within a 5-10 acre range.

Black-tailed Jackrabbit: *Lepus californicus*

The black-tailed jackrabbit is a large hare species native to the western United States and parts of Mexico. Recognized by its long ears and distinct black tail they are built for speed and endurance, with powerful hind legs enabling bursts up to 40 mph (64 km/h). Their long ears help dissipate heat in hot climates, and their coat blends with dry vegetation, providing camouflage.

This species thrives in open landscapes with sparse vegetation, avoiding dense forests. It occupies deserts, sagebrush plains, and grasslands.

Primarily nocturnal and crepuscular, black-tailed jackrabbits forage at dawn and dusk to avoid heat and predators. They are solitary but may congregate where food is abundant.

Courtship often involves chasing behavior, where males pursue females across open ground. Females give birth in shallow depressions called forms, usually hidden in grass or under shrubs. They are born fully furred, with open eyes, and are able to move shortly after birth. They remain hidden and still for much of the day, relying on camouflage rather than a nest for protection.

Population fluctuations of the species often mirror predator cycles, reflecting their integral role in arid ecosystem balance.

84

Antelope Jackrabbit: *Lepus alleni*

The antelope jackrabbit lives in southern Arizona and northern Mexico. It prefers desert grasslands and open plains. It depends on sparse shrubs for shade. It is adapted to hot climates.

Breeding often depends on seasonal rains. Young are born fully furred and active. Growth is rapid during favorable conditions. Survival is tied to rainfall patterns.

This hare has extremely large ears. These ears help regulate body temperature. It is one of the most heat-tolerant North American hares. Its long legs allow for speed and a quick escape.

Photo credit:
Juan Cruzado Cortés

Note the damaged ear.
Photo credit:
Francisco Farriols Sarabia

Ways To Keep In The Know!

YouTube @foxruneec

Join my YouTube channel

Website: www.foxruneec.org

Thank You for your support!

Ame

Biography

Ame Vanorio is an organic gardener, environmental educator, and wildlife rehabilitator. She was raised on a traditional Kentucky farm with horses, cattle, and tobacco. When not pulling weeds she was sneaking off to the back fields to sit quietly communing with wildlife. She rescued numerous baby rabbits, squirrels, and birds that were orphaned or injured due to farm activities like tree work & mowing.

Ame's hands-on experience includes:

- 30 years in education, licensed in environmental science, biology, & special education
- 17 years working in wildlife conservation and rehabilitation. I worked for Nation Wildlife Federation & the Cincinnati Museum of Natural history and for 12 years had my own rehabilitation center.
- Volunteering with Audubon, Sierra, club as well as my own Fox Run EEC to monitor wildlife and teach free environmental education to minimalized populations.

Ame holds graduate degrees in Education and Environmental Science and is the Founder/Director of Fox Run Environmental Education Center.

Check Out My Amazon Author Page - Take a pic of the QR Code below

Resources

Where to purchase supply's

Amazon

Amazon offers a great selection of pet supplies and can be a good source for hard to find things. I get my enclosures through them. Please go to my website and follow any Amazon link to their page. I may receive a small (1-3%) commission on purchases which helps us out in supporting new wildlife rehabilitators. If you are a licensed wildlife rehabilitator Amazon is great for making a wish list and sharing it with your followers.

Henry's Pets

Online pet store that specializes in products geared toward raising small pets.   They carry a line of rehabilitation supplies such as heating pads, formula, supplements, and pest treatments. Discounts for licensed wildlife rehabilitators. I've ordered from them several times and had excellent service.
www.henryspets.com

Exotic Nutrition

Online pet store for exotic pets. They have some good enclosure options, toys and climbing apparatus, and a nice nesting box. They also carry a variety of foods for exotic pets but avoid the treat foods.
https://exoticnutrition.com/

Locally

Many things on my medical cabinet checklist (Free Download on my website) are things you can get locally either at a big box store, feedstore, or your local family pharmacy. In fact I encourage you to develop local relationships as they can be great about giving donations to local charities. If you do not have a garden a great way to get fresh produce for your animals is at a local farmers market

Resources cont.

International Wildlife Rehabilitation Council

They offer a monthly free webinar (Coffee and Tea) and several high-quality courses. Also their course "The Basics" is required by many states for licensing. Courses have In-Person and Online options. If courses are beyond your budget email them and ask if any scholarship monies are available. Sometimes they have assistance. Membership includes their journal and course discounts. https://theiwrc.org/

They publish along with the National Wildlife Rehabilitation Association **The Minimum Standards for Wildlife Rehabilitation, 4th edition, 2012** available on the website.

Fox Run Environmental Education

As I am able to raise funds I am gifting new wildlife rehabilitators an enclosure or sending formula or helping on a Baby Warm incubator. Application process. Funds are limited but see the website for more information. I'm also doing some online information sessions.

America Veterinary Medical Association (AVMA)

Great website with lots of resources however some information is behind an expensive membership pay wall. https://www.avma.org/

Cornell Wildlife Health Lab

Lots of information and resources on their website. https://cwhl.vet.cornell.edu/

National Wildlife Rehabilitators Association

Resources, courses, and an annual conference. https://www.nwrawildlife.org/

References

BSAVA Manual of Rabbit Medicine (BSAVA British Small Animal Veterinary Association) 1st Edition
by Anna Meredith (Editor), Brigitte Lord (Editor)
Amazon https://amzn.to/4qEwIpX

Cornell Wildlife Health Lab (2026). Rabbit Hemorrhagic Disease Virus. NYS Wildlife Health Program.
https://cwhl.vet.cornell.edu/resource/rabbit-hemorrhagic-disease-virus.

Davies, R. et al. (2003)
Rabbit gastrointestinal physiology
Veterinary Clinics: Exotic Animal Practice, Volume 6, Issue 1, 139 - 153

Hentz, P. (2023)
Reducing Anxiety in Eastern Cottontail Rabbits
https://ncwildliferehab.org/wp-content/uploads/2023/01/Reducing-Anxiety-in-Eastern-Cottontail-Rabbits-Hentz.pdf

Marissa Jacky
Raising Rabbits for Facilities
Wildlife Rehabilitation Technician II, Mammal Lead SPCA for Monterey County
https://static1.squarespace.com/static/5cc0813da56827ea5d68b7c4/t/5dead77f20f812386615d742/1575671721632/Raising+Rabbits.pdf

Merck Veterinary Manual
Being older I still use the print manual! However the website has a lot of high quality articles with images. This is made for domestic animals but much of the information is transferable and they are increasingly referencing exotic pets.
https://www.merckvetmanual.com/

 https://www.msdvetmanual.com/exotic-and-laboratory-animals/rabbits/clinical-examination-of-rabbits
https://www.msdvetmanual.com/exotic-and-laboratory-animals/rabbits/nutrition-of-rabbits
https://www.msdvetmanual.com/exotic-and-laboratory-animals/rabbits/noninfectious-diseases-of-rabbits

Ann Mizoguchi
Care and Release of Black Tailed Jackrabbits
https://www.mendowildlife.com/wp-content/uploads/2016/08/Jackrabbit-Care-Release_March-2015.pdf

van der Sluis M, van Zeeland YRA, de Greef KH.
Digestive problems in rabbit production: moving in the wrong direction?
Front Vet Sci. 2024 Feb 7;11:1354651. doi: 10.3389/fvets.2024.1354651. PMID: 38384954; PMCID: PMC10879550.

Willette, M., Rosenhagen, N., Buhl, G., Innis, C., & Boehm, J. (2023).
Interrupted Lives: Welfare Considerations in Wildlife Rehabilitation. Animals, 13(11), 1836. https://doi.org/10.3390/ani13111836